Project Kickoff

How to Run a Successful Project Kickoff Meeting in Easy Steps (Includes a Free Project Kickoff Meeting Template and Free Project Kickoff Meeting Agenda Template)

By

Hassan Osman

Project Kickoff: How to Run a Successful Project Kickoff Meeting in Easy Steps
Copyright © 2019 by Hassan Osman.

Notice of Rights
All rights reserved. No part of this publication may be reproduced, distributed, or transmitted in any form or by any means without the prior written permission of the author. Reproduction or translation of this work in any form beyond that permitted by section 107 or 108 of the 1976 United States Copyright Act is strictly prohibited. For permission requests, please contact the author. Reviewers may quote brief passages in reviews.

Liability Disclaimer and FTC Notice
The purpose of this book is to provide the user with general information about the subject matter presented. This book is for entertainment purposes only. This book is not intended, nor should the user consider it, to be legal advice for a specific situation. The author, company, and publisher make no representations or warranties with respect to the accuracy, fitness, completeness, or applicability of the contents of this book. They disclaim any merchantability, fitness or warranties, whether expressed or implied. The author, company, and publisher shall in no event

be held liable for any loss or other damages, including but not limited to special, incidental, consequential, or other damages. This disclaimer applies to any damages by any failure of performance, error, omission, interruption, deletion, defect, delay in operation or transmission, computer malware, communication line failure, theft or destruction or unauthorized access to, or use of record, whether for breach of contract, tort, negligence, or under any other cause of action.

By reading this book, you agree that the use of it is entirely at your own risk and that you are solely responsible for your use of the contents. The advice of a competent legal counsel (or any other professional) should be sought. The author, company, and publisher do not warrant the performance, effectiveness or applicability of any sites or references listed in this book. Some links are affiliate links. This means that if you decide to make a purchase after clicking on some links, the author will make a commission. All references and links are for information purposes only and are not warranted for content, accuracy or any other implied or explicit purpose.

Table of Contents

Introduction 5

Your Free Bonus 12

Section I: Before Your Project Kickoff Meeting 13

Section II: During Your Project Kickoff Meeting 21

Section III: After Your Project Kickoff Meeting 49

Conclusion 54

Thank You! 56

Introduction

Every successful project needs a good kickoff meeting.

As anyone will tell you, first impressions are everything, so making a good one is vital. In the world of project management, that chance comes during the project kickoff meeting, which signifies the official start of your project.

If you leave a good first impression on your attendees, then you'll set the stage for a successful engagement. If, on the other hand, your first impression is less than stellar, you might leave a negative perception that will be hard to recover from later on. It's therefore essential to prepare well for your project kickoff and make sure you have a structured method of approaching it.

An effective project kickoff meeting accomplishes three things:

First, it jumpstarts the project by formally bringing together the relevant stakeholders and informing them of the project goals.

Second, it introduces the different players to each other so that they're aware of everyone's roles and responsibilities.

And third, it generates enthusiasm among the team members, motivating everyone to get started on the project.

In this book, I'll give you a simple framework to help you accomplish all these goals. It's a short read that will show you how to run a successful project kickoff meeting in simple steps.

Let's get started.

Who is this book for?

This book is for project managers who lead project teams in medium to large-sized organizations. It's also for business owners and entrepreneurs who work on project-based work.

The book is intentionally short because it's boiled down to its essentials. I don't want to waste your time with useless padding, so I removed any fluff that didn't add much value. But don't equate the brevity of this book with a lack of usefulness. You'll get a clear, step-by-step process to run a successful project kickoff meeting.

Why should you listen to what I have to say?

I manage projects and project management teams for a living.

I'm currently a Project Management Office (PMO) leader at Cisco Systems, where I lead a team of over 150 project and program managers on delivering complex projects across the US and Canada (lawyer-required note: the opinions in this book are mine and not those of Cisco).

Prior to Cisco, I was a management consultant at Ernst & Young (now EY), where I ran projects and programs at Fortune 100 companies.

I'm a certified PMP (Project Management Professional) and CSM (Certified ScrumMaster), and I've managed or supervised hundreds of projects throughout my career.

Every single project I oversaw required a project kickoff meeting, and I have experienced firsthand what works and what doesn't during project launches.

This book contains the best of the best practical advice that will help you run a successful project kickoff meeting yourself.

You'll get simple tactics that you can implement straight away to help you prepare, as well as scripts that you can copy and paste to save you time.

How is this book organized?

This book is organized into three main sections. Section I is about what you need to do *before* your kickoff meeting. Section II covers what you need to do *during* your kickoff meeting, and Section III is about what you need to do *after* your kickoff meeting.

In each section, I cover the exact steps you need to take to ensure that your meeting is successful.

Although some of these steps might seem elementary or redundant, don't skip over any of them. They have been carefully thought out as part of a comprehensive end-to-end process.

Quick definitions

I'm not a big fan of getting bogged down in definitions, but it's important to clarify a few terms before we start to clear up any confusion.

- **Project kickoff meeting**: A project kickoff meeting is the first meeting that takes place between the *project team* and the *project customer*. It is typically conducted after a project has been scoped, funded, and approved.

- **Project team**: A project team is a group of individuals who are responsible for delivering a project.

- **Project manager**: The project manager is the leader of the project team and is responsible for managing and running the project kickoff meeting (note: in this book, I assume you are the project manager).

- **Project customer**: The project customer is the individual, team, or organization that needs the project.

- **Project sponsor**: The project sponsor is an individual who is part of the project customer and is the main person accountable for the project. They usually secure funding for the project.

- **Sales team**: The sales team is the group of individuals that sold the project. In some organizations, the sales team might be part of the project team (someone like a sales account manager could be involved in the day-to-day management of the project). In other organizations, the sales team is separate from the project team.

- **Stakeholders**: The stakeholders are all the members involved in a project (including the project team, project manager, project customer, project sponsor, and sales team), as well as any other individuals who are affected by the project's outcome.

A note about internal vs. external

Throughout the book, I refer to *internal* and *external* teams or meetings.

For simplicity, I assume that you are delivering a business-to-business (B2B) kickoff meeting. In other words, I assume you're part of a team in one organization that is delivering a kickoff meeting to a team in another organization.

The word *internal* refers to anything related to your internal organization (i.e., the organization delivering the kickoff meeting), and the word *external* refers to anything related to the customer organization (i.e., the organization on the receiving end of the kickoff meeting).

This means that the project manager, project team, and sales team are all part of the internal organization. And the project customer and project sponsor are part of the external organization.

Note: If you're delivering a project to another department or business unit within your organization, then the same steps for delivering a project kickoff meeting apply. Just assume that the other department or business unit is the project customer.

With that, let's get started with what you need to do before your kickoff meeting.

Your Free Bonus

As a thank you for your purchase, I'm offering a free bonus that is exclusive to my readers.

This bonus includes a couple of templates to help you with your kickoff meeting: a **project kickoff meeting template** and a **project kickoff meeting agenda template** that you can use with your own team.

The templates match the guidelines in this book, so you will save time in creating your own documents. I'll also explain how to use these templates throughout the book.

They are in Microsoft Word (.doc) and Microsoft PowerPoint (.ppt) format so that you can start using them right away.

Visit the following page to download your free bonus:

www.thecouchmanager.com/pkbonus

Section I: Before Your Project Kickoff Meeting

In this section, we'll cover the five steps that you need to take before your project kickoff meeting. These include: drafting the agenda, gathering feedback, preparing the presentation, conducting an internal kickoff meeting, and scheduling the external kickoff meeting.

Step One: Draft the Agenda

The agenda sets the structure for the entire kickoff meeting. It defines what people will talk about and in what order.

A robust agenda starts with a high-level objective statement that clarifies its purpose.

For example:

"The objective of this kickoff meeting is to bring together all the key stakeholders so that we can <insert your high-level project goal here>."

Then, the agenda should list the key topics that the meeting will cover, such as:

- Welcome
- Introductions
- Project sponsor update
- Project goals
- Team organization chart
- High-level schedule
- Project assumptions and constraints
- Communication plan
- Milestone signoff and invoicing process
- Technical update
- Other considerations
- Next steps
- Q&A session
- Wrap up

These topics are usually a part of every successful project kickoff meeting, and we'll discuss each topic in detail in *Section II: During Your Kickoff Meeting*.

Part of creating the agenda also involves setting the total length of the meeting. In large organizations, scheduling at least two hours should be enough.

However, it's a good idea to allocate three to four hours for the meeting to ensure time for all stakeholders to air their views and seek clarifications.

You should also set time limits for every topic to avoid using up much-needed time. Make sure that your agenda indicates how much time every presenter will have.

The output of this step is a *preliminary* draft of your agenda with assigned presenters and allocation of time for each presenter. Keep in mind that it's normal for this draft to be modified a few times after you share it with other team members, so it's not set in stone at this point.

Step Two: Gather Feedback

Once you have drafted the agenda, the next step is to gather feedback about it from your stakeholders.

Your objective here is to accomplish two things. First, you're seeking input from key stakeholders about any improvements to the agenda. And second, you want to confirm the logistics of the meeting and identify who will attend.

Here are a few questions that you might ask when seeking feedback:

- What are some potential additions or modifications to the agenda topics?
- How large is the available conference room?
- Who needs to attend?
- What collaboration or meeting tool are we going to use? (WebEx, GoToMeeting, etc.)
- What time or date works for the team?
- How many people will be attending in-person versus remotely?
- How would you make this agenda better?

The answers to all these questions could affect how your agenda gets modified because different projects have different constraints. For example, the project customer might state that they only have an hour to spare and a conference room that fits a maximum of ten people, which means you'll have to cut down on specific topics and ask any prospective in-person attendees to attend remotely instead.

A best practice is to start gathering feedback from your *internal* stakeholders first—such as your project team and sales team—because you might get some ideas about modifying the agenda topics before you share them with the project customer.

For instance, the sales team might suggest that you should avoid giving a technical update during the kickoff meeting because most of the customer team members don't have a technical background.

The output of this step is an updated agenda that is modified based on potential constraints about logistics (e.g., meeting time, meeting duration, number of attendees, etc.), as well as suggestions for improvement.

Step Three: Prepare the Presentation

After updating the agenda and solidifying your outline, your next step is to prepare the presentation deck of your kickoff meeting. This step involves consolidating all the different sections that will form your project kickoff slides.

The most popular presentation tool that organizations use is Microsoft PowerPoint, but other useful applications include Keynote, Prezi, and Google Slides.

Expect to spend a significant amount of

time on this step as you put together different sections from different stakeholders. You might need to spend a few days—in some cases weeks—to get this complete. It's not uncommon to gather some information from the customer during this step, such as information about their team structure or roles, to add to the slides.

As you go through the process, make sure the flow of the presentation matches the agenda outline.

As a reminder, I have included a free downloadable file that will help you save time in creating your presentation. It also contains examples of each section to give you an idea of what you should cover. If you didn't download the file earlier, visit the following link to download it:

www.thecouchmanager.com/pkbonus

Step Four: Hold an Internal Kickoff Meeting

After the presentation is complete, your next step is to hold an internal kickoff meeting with your team.

This meeting is a dry-run of the actual meeting to review the presentation that you will deliver later to the customer. The objective of this practice run is to ensure that your internal team—including the project team and sales team—are all on the same page before your external kickoff meeting. It'll also help you and other presenters practice delivering the meeting ahead of time with a real audience.

Moreover, it'll also help the team test the technology by practicing the use of any remote collaboration tools. Doing so will make sure that all presenters are comfortable with the flow, especially if there are any transitions between speakers.

Based on feedback from your team during your internal kickoff meeting, you might need to make some changes to the final presentation deck.

If these modifications are minor, then you can make the changes and move on to the next step.

However, if the changes are substantial, then you might need to schedule another practice internal kickoff meeting. That's because you want to make sure that everyone is fully aligned with those significant changes before you meet with the customer.

Step Five: Schedule the External Kickoff Meeting

After holding a successful internal kickoff meeting, your next step is to schedule the external kickoff meeting with the customer.

And when you send the invite to the meeting, remember to attach the agenda so that everyone is aware of what to expect. Moreover, if there are any materials or documents that attendees need to review before the meeting, then send all that information along with the invite as well so that everyone comes prepared.

It's also a good practice to send a courtesy reminder about the meeting a few days before the meeting date.

After scheduling the meeting, your next step is to deliver it on the scheduled date.

Section II: During Your Project Kickoff Meeting

In this section, we'll cover the steps that you need to take *during* your project kickoff meeting. This includes starting with a welcome and introduction, getting an update from the project sponsor, and then walking through the entire presentation before closing the meeting out with a Q&A session and wrapping it up. There are also a few steps that I consider optional depending on your project (and I've marked them as such below), so use what you need.

One recommendation I have while you're presenting the content is to assign someone from your internal team to take notes during the meeting. Doing so helps you as a project manager focus on delivering and facilitating the kickoff meeting without having to worry about taking meeting minutes or getting distracted.

Step One: Welcome and Rundown of Agenda

When you start a project kickoff meeting,

your welcoming remarks should be enthusiastic, warm, and motivating. Remember that you're bringing together cross-functional team members that may have never worked together before, and the impression you create will set the tone for how the rest of the project will follow. Focus on projecting an image of confidence and control as you kickstart your meeting.

Even if the project is about fixing a major problem, emphasize the positive aspects and don't let negativity dampen the mood. Highlight the fact that everyone is in one room to come up with a solution.

Here is a sample script that you can use as your opening speech:

"Hello and welcome to the <project name> kickoff meeting. My name is Hassan Osman, and I'll be facilitating the meeting today. I hope everyone is doing well, and it's great to see you all here together. We have assembled a talented group of professionals who can tackle this project head on to meet our goals and objectives.

<Add the following if the project is meant to fix a problematic issue>

I realize that we are meeting under somewhat unpleasant circumstances, but I want you to know that if we pull together, we can resolve this issue as quickly and effectively as possible. Thank you all for being a part of this team."

In case some or all the attendees are attending the meeting remotely, take some time to acknowledge them and make sure that everyone can hear and see what you are presenting.

For example:

"I would also like to welcome those of you who are joining us remotely. Are you all able to see the presentation slides? <Pause to check for confirmation> Great! Let's move on."

After welcoming the attendees, give a brief rundown of the agenda items. This is important because people always prefer to know what to expect during a project kickoff meeting. You should also ask people to reserve any questions they have until the end of the session.

For example:

This project is key to the organization because <insert relevant benefits>.

And here's what we'll be covering today:

<Provide a rundown of your agenda items>."

Feel free to modify this script accordingly. The important thing to keep in mind is to get people fired up for the meeting and to create a sense of camaraderie among the stakeholders.

Step Two: Introductions

Introductions are essential because they help break the ice and ensure that everyone in attendance feels that they are part of the team. After welcoming everybody and running through the meeting agenda, the next step is to ask all the attendees to introduce themselves briefly.

The keyword here is "briefly" to avoid having folks eat up half the meeting talking about themselves. Make sure that you set clear expectations about what you want them to say. You do not want to have a situation where someone uses up several minutes blabbing on and on (it happens more frequently than you think). Just ask them to take a few seconds to tell their colleagues who they are, their role in the project, and the organization they work for.

It may seem unnecessary to get people to

state their organization, but consider that sometimes it is difficult for people to be sure of who works for who, especially when large organizations are collaborating on a project. I've led several kickoff meetings where members of the project customer had never met before and didn't even know they worked at the same company!

As the person who is leading the kickoff meeting, you should be the first to introduce yourself. Doing so will also help set the expectation because people will follow your example. The attendees who are physically present should introduce themselves next, followed by those who are joining you remotely. It is easy to overlook the people who are dialing in, so don't forget to ask them.

Here is a sample script that you can use:

"Let's go around the room for some quick introductions. Please state your name, the organization you work for, and the role you're playing on this project. Unfortunately, we'll need to keep these intros a bit brief given that this is such a large group.

Let me begin by introducing myself. Hello, my name is Hassan Osman. I work for Cisco Systems, and I'll be the lead project manager on this project. I'll let the project sponsor and customer team introduce themselves, and then the rest can follow."

Step Three: Project Sponsor Update

After going through the introductions, set aside some time for the project sponsor to provide the team with an update. The project sponsor, as the leader from the customer side, can use this opportunity to explain to the attendees why the project is significant. They can also give a high-level vision of what they expect at the end of the project.

If relevant, the sponsor can also provide the team with some history behind the project and how it fits into the overall strategic plan.

The primary objective here is to leverage the project sponsor's presence to motivate and inspire the team.

If the project sponsor is not available to attend, then an alternative is to use a quote or some thoughts that are shared by the sponsor ahead of time. This will have a similar motivating effect on the team.

Step Four: Project Goals

After the project sponsor update, the next step is to discuss the project goals and objectives. Here is where you'll review the deliverables of the project and the measures of success.

Start by outlining the goals and objectives that are listed in the contractual documents. If there's a signed Statement of Work (SOW), a project charter, or another official document that has been agreed to by both sides, then use that as the source to extract the information you need for this section.

It's always a good idea to use the *exact* verbiage used in contractual documents for a couple of reasons. First, you'll avoid any potential miscommunication issues later because modifying the wording could inadvertently change the original intent. And second, you'll get the chance to remind the team of what was already agreed upon before the project starts.

You don't have to provide all the details listed in the contract (SOWs can be very lengthy documents), but do highlight the key takeaways.

Another thing to include is the output of your project: your deliverables or work products. These are the tangible products which could include elements such as a software package, a physical product, or a set of documents. Again, it would be wise to add the exact wording used to describe the deliverables in your contractual documents. That's because things can easily get lost in translation if you modify the original text.

To help level-set on expectations, list out all the deliverable names, as well as a description of each. I like to use a small table here to explain what those work products are.

Finally, include a section about how the team will define success. This will be a value statement that focuses on the outcome of the project. Defining success will be different from one project to another, but the idea here is to think beyond just delivering on the agreed upon scope, budget, or time, and instead considering what the customer is hoping to achieve in the long run.

For example: *"Success on this project means that we will have 50,000 users fully onboarded on this new system by May 13,* *which should increase productivity by 35%. This should save $2.7M in the first year after implementation."*

Your goal in this step is to show the project customer what they are paying for so that you explain the value you're adding to their business. This step will also give you the opportunity to uncover any misaligned expectations, and to address them early on.

Step Five: Team Organization Chart

After the project goals, the next step is to share the team organization chart.

The organization chart illustrates the hierarchy within the project team and the roles that will be played by the different team members. It will also show how the team will work together.

A best practice is to show the pictures, names, and roles of each project team member, along with their reporting lines.

Keep in mind that this should be presented from a *customer-facing* view and that any roles and relationships should be modified to that view.

For example, if an engineer on your project team will report to you as a project manager on the project, but reports to a different manager within your internal organization, then only show the former relationship on the organization chart.

Another example is if someone's role on your project is a "Change Control Lead," but their official title in your internal organization is a "Business Analyst," then only show the former role on the organization chart.

The customer usually doesn't care about your internal titles or structure, so only present the roles that matter.

Depending on how large the team is, you might want to illustrate how the team members fit into different functions. For example, you can organize your team into a "sales function," and an "engineering function." You can also organize by tracks or workstreams within the project, such as a "strategy track," a "technology track," and a "finance track."

Do what makes sense to your team and project.

In some engagements, the project customer might have their own team that is actively involved in the project. In that case, it's also a good idea to show a separate slide with the customer's organizational structure. Doing so can highlight who are the key players on the customer side while also avoiding confusion about who works where.

Note: Some project managers also like to include a "roles and responsibilities" table showing details on what each role will do, as well as a governance chart that shows how the different resources will be governed. I think that these two elements are unnecessary at the kickoff meeting and that the team organization chart(s) should suffice to explain those roles and relationships. You can always share more detailed information in subsequent meetings.

Step Six: High-Level Schedule

After presenting the organization chart, the next step is to show a high-level schedule.

This schedule should be a simple timeline that displays the significant milestones and phases of a project, and when they're expected to be complete.

While some project managers like to share their Gantt chart—a detailed bar chart showing all activities and dependencies—during the kickoff meeting, that level of detail might be too much for a kickoff audience.

My preference is to keep the schedule at a much higher level, where you focus on the major milestones and phases. A useful heuristic to follow is that if the timeline doesn't fit legibly with clear dates and milestone labels on *one* slide, then modify it so that it does (see the template for a good example).

Another reason to keep the schedule at a high level and not go into too much detail is to give your project team some leeway with potential future changes. In almost every project, things don't go as planned, and you will gain more information and clarity as you execute on your project. One idea is to add a footnote on your schedule slide to mention that this a preliminary draft only, and that you'll share a more updated version of the schedule as the project proceeds.

Doing so also helps avoid being held accountable for failing to meet milestones or deadlines simply because the customer was referring to an old version of the schedule.

An effective schedule serves two purposes. First, it provides a sense of situational awareness about the sequence of activities and their relationship. The majority of people are visual, and looking at a graphic helps them see how the project pieces fit together.

Second, it helps set expectations with your stakeholders, particularly the project customer, about reasonable deadlines. Most customers want things done quickly, and showing a schedule that helps identify how the dependencies and roadblocks affect your dates will assist in level-setting their expectations.

Step Seven: Project Assumptions and Constraints

After presenting the high-level schedule, the next step is to list out any significant project assumptions or constraints.

These are factors that can affect your project's scope, schedule, or budget, and it is important to mention them in your kickoff meeting. To reiterate, it's a good idea to use the same language you used in your contractual documents (like a SOW) to describe project assumptions and constraints.

Assumptions

Assumptions are factors that are expected to be true in your project.

Here are a few examples:

- **Scope boundaries**: Include any major out-of-scope elements, such as activities, products, or services that are not part of the original project plan.

- **Frameworks or Methodologies**: List out any frameworks or methodologies that you'll be using throughout the project. For example, explain whether you'll be following an agile or waterfall methodology. You can also mention any standards or frameworks you'll be using such as PMI, Scrum, Kanban, or PRINCE2.

- **Technology Tools:** Highlight the different tools you'll be using throughout the project, including applications for collaboration, scheduling, and document sharing (e.g., WebEx, Basecamp, Box, etc.). And if your project team will need access to the customer's environment (laptops, online accounts, etc.), mention these assumptions here, as well.

Constraints

Constraints are factors that limit your ability to meet the expected results in your project.

Here are a few examples:

- **Risks:** Mention any significant risks that will affect your project, and briefly discuss any contingencies that you have put in place. A detailed risk register should be part of your core project management documents, but you don't want to get into that level of detail here. Keep it brief and address any risks from a general perspective.

- **Dependencies:** List out any key dependencies that affect the successful completion of the project. This could include bottlenecks or tasks that should be completed by a specific deadline before others can proceed. Other examples include things like change freezes if you're conducting any technology upgrades.

- **Resources/Budget**: If you have any limitations on your resources, such as staff or equipment to get your project complete, then list them here. Similarly, if there are any budget limits on certain aspects of the project, you should highlight them here as well. For example, you might have a $3,000 budget and 50% cap on a developer's time for a specific activity that you want to point out as a limiting factor.

The objective in this section is to focus on the *major* assumptions and constraints with your team, so keep things concise.

Step Eight: Communication Plan

After reviewing the project constraints and assumptions, the next step is to discuss your communication plan.

A communication plan is an absolute-must component of your kickoff meeting because attendees want to know what to expect regarding how and when you're going to communicate with them.

Although communication plans can be pretty comprehensive and would probably require a deep-dive in a separate meeting, you should cover at least three main components during your kickoff: Meetings, Contact Information, and Escalation Path.

Meetings

List out all the different types of meetings you'll have with the team, such as status meetings, technical meetings, and senior management meetings, as well as their frequency (e.g., weekly, biweekly, monthly, etc.).

You should also include the audience members who will join each meeting and the communication tools you'll use. A best practice is also to include the output of each meeting, such as the reports that your team will generate (e.g., status reports, technical charts, or meeting minutes).

Contact Information

List the contact information of all your project team members. Include their roles, email addresses, preferred phone numbers, and any other relevant information for the project customer to get in touch with them. Having all that information on one slide will come in handy later.

Escalation Path (optional)

It is also a good idea to show the project team an Escalation Path that will guide them on who to go to in case of any issues.

If the size of your team is relatively small, then the escalation path might be obvious, and you would not need to show it.

If you decide to include one, here are a couple of examples that illustrate what it looks like:

Engineer → Project Manager → VP of Services

OR

Project Manager → IT Sponsor → Business Sponsor → Program Executive

You can also list out escalation paths that are based on the issues your customer might face. For example, you can have one escalation path for technical issues and another escalation path for resource issues.

I recommend having a separate slide for each one of these three components (Meetings, Contact Information, and Escalation Path) in your communication plan, and I have included examples for you in the template I shared earlier.

Step Nine: Change Management Process

After reviewing the communication plan, the next step is to discuss your change management process.

A change management process covers how you're going to address unforeseen changes that pop up during your project. For example, if the customer decides to increase the scope of the project after you start, or a new deadline is introduced that accelerates your schedule (and, consequently, increases your cost), then the change management process is what you will follow.

Although some project managers leave this portion out during the kickoff, I think it's a huge mistake. Unplanned changes are almost guaranteed to occur, and you want to set expectations up front about how you'll handle these changes.

A change management process should include clearly defined steps about how to request a change, and who will approve it among the stakeholders. If you have a standard "Change Request" (CR) template, then present a sample of it during your meeting so that everyone gets the chance to see it.

Setting the expectation that any project changes will require a formal process will help you mitigate any awkward or uncomfortable conversations with the customer down the line. This is especially true if you must ask for more funds due to an increase in scope.

Step Ten: Milestone Signoff and Invoicing Process (optional)

After the change management process, your next step is to present the milestone signoff and invoicing process, which covers how to get paid for your services.

Most projects are paid on a milestone basis—at successive points in time after significant portions of the project are complete.

This step sets clear expectations on how the work will be reviewed, accepted, and invoiced after completion. A good idea here is to show a table with all the different milestones, deliverable names, and payment amounts associated with each milestone. This information is usually listed in your contractual documents.

You should also show the different stages of deliverable reviews, the expected timeframes, and who will have the final say on signing off at completion. This is usually the project sponsor, but they might defer to a senior leader on their team.

If you have a standard milestone completion form, then present a sample of it in the meeting so that everyone is familiar with it. Also, if you have an acceptance policy based on non-responsiveness, then state it here, as well. For example, some companies require that customers sign off on a milestone within five business days of being sent the milestone completion form. And if no response is registered (if, for example, someone is out of the office), then your organization has the right to deem the work product as accepted and to send an invoice for that specific milestone.

Sharing the milestone signoff and invoicing process during the kickoff meeting is optional because, depending on the audience, bringing up the topic of payments can be sensitive or irrelevant. However, if you decide not to share it here, I strongly recommend that you address it in a separate meeting with the project sponsor and any other stakeholders. You want to manage any risks that might surface as a result of the discussion around payments.

Step Eleven: Technical Update (optional)

A technical update is not mandatory in your kickoff meeting, but, in some instances, it may be a good idea to dedicate some time to highlight the technical aspects of your project.

The technical update should focus on discussing the content of your major technical deliverables.

For example, you can allow your lead engineer a few minutes to discuss the methodology they will use to fix a particular technical problem. Or you can give your product developer an opportunity to provide an overview of how they intend to approach the design.

Including a technical update can be an effective way for the rest of the stakeholders to gain some insight into the dynamics of the project. This is particularly useful when you are dealing with very technical projects that are designed to solve complex problems.

One thing to note is to keep this section brief and avoid slipping into a deep-dive conversation.

Step Twelve: Other Considerations (optional)

After the technical update, the next step is to address any other considerations that may not necessarily fall into the previous categories.

This is sometimes referred to as the AOB (*Any Other Business*) section, and it can cover anything you want to address while you have all the relevant stakeholders present in the meeting.

For example, you can mention how you

intend to monitor your project's progress and budget. Or you can discuss how you will comply with corporate policies that affect your project.

This is an optional section that you can leave out if everything you need to cover is already covered in the previous sections.

Step Thirteen: Next Steps

As you get close to the end of your meeting, the next topic to address is your next steps section.

That's because everyone naturally wants to know "where do we go from here and what should we do next?"

So be proactive and state all the relevant next steps. For example, you might want to mention that you'll gather project documentation, schedule interviews for data analysis, and set up weekly status calls with the project team.

A best practice is to also include owners for each action item and the dates that they're due. This will ensure that the project continues on the right path as you drive the team forward.

If any other actions popped up during the meeting, then verbally address those here, as well. This will give stakeholders the confidence that you were listening and that you will act on those tasks.

Step Fourteen: Q&A Session

After covering the next steps, your next topic is a question and answer session.

A Q&A is an open forum for stakeholders to ask questions and seek clarification on different subjects. The advantage of scheduling a Q&A session at the end of the meeting is that it gives you more control over the conversation during the meeting.

For example, if someone interrupts you with off-topic questions during your presentation, you can politely ask them to wait until the Q&A session to address them. Doing so will help you stay on track and cover everything you want to go over before you run out of time.

During the Q&A session, take note of all the questions and answers that come up so that they're part of your meeting minutes. And if there are any questions that you cannot answer, or you run out of time, then let everyone know you'll address them in a follow-up meeting.

One thing to remember is to allow those who are following remotely to also engage in the Q&A discussion. Since they are not physically in the room, it is easy to forget that they're present.

Step Fifteen: Wrap Up

After the Q&A session, your last and final step is to wrap up your project kickoff meeting. Always leave a buffer of a few minutes before your scheduled end time to close out the meeting with a couple of statements.

There's a high chance that you'll still have a few unanswered questions from the Q&A session. In that case, interject with a respectful comment that you need to wrap the meeting up and that you'll address all their questions later. This shows that you know how to control a team meeting and that you respect everyone else's time.

As you wrap up, thank everyone for their time, and state that you had a successful meeting. You want to close out on a high note, regardless of how things materialized. So even if things unexpectedly got a bit heated or contentious during the meeting, acknowledge that there were some challenges but that you'll solve them as a team. The key is to wrap it up with enthusiasm and a positive vibe.

Here's an example:

"Thank you all for taking the time to attend the meeting today, including folks who dialed in remotely. I think we had a very successful kickoff and I'm really excited about our next steps to get this moving. I know there were a couple of opposing views that came up, but we have a stellar team, and I'm sure we'll iron those issues out as we progress."

The idea is to focus on something constructive to say and to prevent anyone from leaving with a negative mood.

If you had planned any informal get-together, such as a dinner or team event, then share the logistical arrangements at this point. It's always a good idea to remind everyone of the place and time the team is going to meet.

After closing out your kickoff meeting, your next step is to follow up with your stakeholders a day or so later.

Section III: After Your Project Kickoff Meeting

In this final section, we'll cover the two steps that you need to take *after* you complete your project kickoff meeting. These include sending an update and following up on your action items.

Step One: Send an Update

A day or two after your meeting, and after you have had the chance to clean up your meeting notes, you should send a comprehensive update to all the stakeholders via email.

Here's what your update should include:

- **A quick summary**: Start your message with a thank you note and a quick summary of the date and time you met, along with a high-level overview of the project objectives.
- **A copy of your presentation**: Attach the project kickoff presentation file with your email as a reference for stakeholders.

- **Meeting minutes**: Also include a copy of your meeting minutes, which should consist of your detailed notes, as well as a description of the actions, issues, and risks that came up during the meeting. If you recorded the session, then include a link to the recording in your meeting minutes as well.
- **Summary of immediate next steps**: Although the project kickoff presentation file and the meeting minutes should already include the next steps that everyone should take, it's a good idea to summarize the immediate next steps in the body of your email so that they're front and center.
- **Relevant links**: If you are using any collaboration, project management, or document management tools in your project, then share the appropriate links to those tools so that your team can start getting familiar with them.

When you send the update out, make sure you also include anyone who couldn't attend the meeting and is a relevant stakeholder. For example, you might want to add the project sponsor's leader or the VP of your internal organization to keep them in the loop on the completion of your kickoff meeting.

Here's a sample script of an update:

Team,

Thank you all for attending the project kickoff meeting on <Date> at <Time>. I'm very excited about the project and the constructive discussion we had. I'm looking forward to bringing the team together to <state project goals>.

Attached is the project kickoff meeting deck that we reviewed during the call.
I have also attached a meeting minutes file, which includes a summary of what was discussed during the meeting, as well as the next steps on our project. The meeting minutes also include a link to the recording of the meeting.

Here's a quick summary of our immediate next steps:

- *<Action>, <Action Owner>, <Due Date>*
- *<Action>, <Action Owner>, <Due Date>*
- *<Action>, <Action Owner>, <Due Date>*

And here are the links to our document repository as well as the collaboration tool. Please make sure you sign up to an account on both platforms as soon as possible so that we start using them as a team.

- *<Collaboration tool link>*
- *<Document repository link>*

Thanks again for your time, and I look forward to working with you all. If you have any questions in the meantime, please feel free to reach out to me directly.

Thank you,

Hassan Osman

Step Two: Follow Up on Action Items

After you send an update, your next and final step is to follow up on all your action items that you highlighted in your kickoff presentation and meeting minutes.

This sounds like I'm stating the obvious, but I want to stress it as a separate next step because some project managers wait for way too long after their project kickoff meeting before executing on their next steps.

Start by scheduling any recurring meetings with the different teams so that you set up your touch points. And if you need to gather any information from some team members before you get started, make sure you obtain those as soon as possible so that you don't run into any delays. Also, schedule any other ad-hoc meetings with a subset of your team if needed, such as requirements-gathering interviews or brainstorming sessions.

The idea behind following up immediately after your kickoff meeting is to continue with the positive energy you established during the meeting so that you get the ball rolling.

Conclusion

We just covered how to run a successful project kickoff meeting in easy steps, and what you need to do before, during, and after the meeting.

Here's a quick summary:

In Section I, we covered the steps you need to take before your kickoff meeting. You should start by drafting the agenda, gathering feedback about it, and then preparing your presentation. After that, you should hold an internal kickoff meeting to review the presentation and logistics, and, once all that's done, you can schedule your external kickoff meeting.

In Section II, we covered the steps you should take during your kickoff meeting. These include starting with a warm welcome, a rundown of the agenda, and a few introductions. Then, after going through the project sponsor update, project goals, and team organization chart, you can review the high-level schedule, project assumptions and constraints, communication plan, and change management process. Optionally, you can also cover the milestone signoff process, technical update, or any other considerations. Finally, you'll end your meeting with the next steps section, a Q&A session, and a quick wrap up.

In Section III, we covered the steps you should take after your kickoff meeting. These include sending an update with your presentation and meeting minutes attached and following up on all the action items that were discussed.

Keep in mind that this end to end process is just a framework intended to help you run your kickoff meeting. However, you can certainly modify it to accommodate your project's needs, so feel free to tweak it as you see fit.

Thank You!

I want to thank you once again for purchasing my book. I hope you found it helpful, and I wish you the best of luck with your kickoff meeting.

I'd like to ask you for a small favor.

If you enjoyed the book, I'd be very grateful if you leave an honest review on Amazon (I read them all).

Every single review counts, and your support really does make a difference.

Thanks again for your kind support!

Cheers,

Hassan

P.S. If you enjoyed this book, you'd probably enjoy my other book entitled "Influencing Virtual Teams."

It's an Amazon #1 Best Seller and covers 17 tactics that get things done with your remote employees.

You can find it by searching for "Influencing Virtual Teams" on Amazon.com

www.ingramcontent.com/pod-product-compliance
Lightning Source LLC
Chambersburg PA
CBHW021512210526
45463CB00002B/986